Rourke Educational Media

rourkeeducationalmedia.com

Animals Have Classes Too!

Mammals

Nancy Furstinger

Before, During, and After Reading Activities

Before Reading: Building Background Knowledge and Academic Vocabulary

"Before Reading" strategies activate prior knowledge and set a purpose for reading. Before reading a book, it is important to tap into what your child or students already know about the topic. This will help them develop their vocabulary and increase their reading comprehension.

Questions and activities to build background knowledge:

1. *Look at the cover of the book. What will this book be about?*
2. *What do you already know about the topic?*
3. *Let's study the Table of Contents. What will you learn about in the book's chapters?*
4. *What would you like to learn about this topic? Do you think you might learn about it from this book? Why or why not?*

Building Academic Vocabulary

Building academic vocabulary is critical to understanding subject content.
Assist your child or students to gain meaning of the following vocabulary words.

Content Area Vocabulary

Read the list. What do these words mean?

- *cells*
- *habitats*
- *marsupials*
- *pores*
- *predators*
- *senses*
- *species*
- *vertebrates*

During Reading: Writing Component

"During Reading" strategies help to make connections, monitor understanding, generate questions, and stay focused.

1. *While reading, write in your reading journal any questions you have or anything you do not understand.*
2. *After completing each chapter, write a summary of the chapter in your reading journal.*
3. *While reading, make connections with the text and write them in your reading journal.*
   a) Text to Self – What does this remind me of in my life? What were my feelings when I read this?
   b) Text to Text – What does this remind me of in another book I've read? How is this different from other books I've read?
   c) Text to World – What does this remind me of in the real world? Have I heard about this before? (News, current events, school, etc....)

After Reading: Comprehension and Extension Activity

"After Reading" strategies provide an opportunity to summarize, question, reflect, discuss, and respond to text. After reading the book, work on the following questions with your child or students to check their level of reading comprehension and content mastery.

1. How do scientists classify mammals? *(Summarize)*
2. What can you conclude about mammals within a phylum as compared to mammals within a class? *(Infer)*
3. How are walruses and dolphins alike? *(Asking Questions)*
4. If you had to make a new way to classify mammals, what traits would you use to sort them? *(Text to Self Connection)*

Extension Activity

Create a set of animal trading cards! Choose three of your favorite mammals to start. Write facts about each mammal on one side of a blank index card. On the other side, draw a picture of the mammal. Trade cards with your friends and classmates.

Table of Contents

mandrill

Let's Classify!

There are so many plants and animals in the world to study! How do scientists keep track? They sort them into groups.

Scientists sort every living thing into kingdoms based on what kind of **cells** it has, how many cells make up its body, and whether or not it makes its own food.

Animal Groups

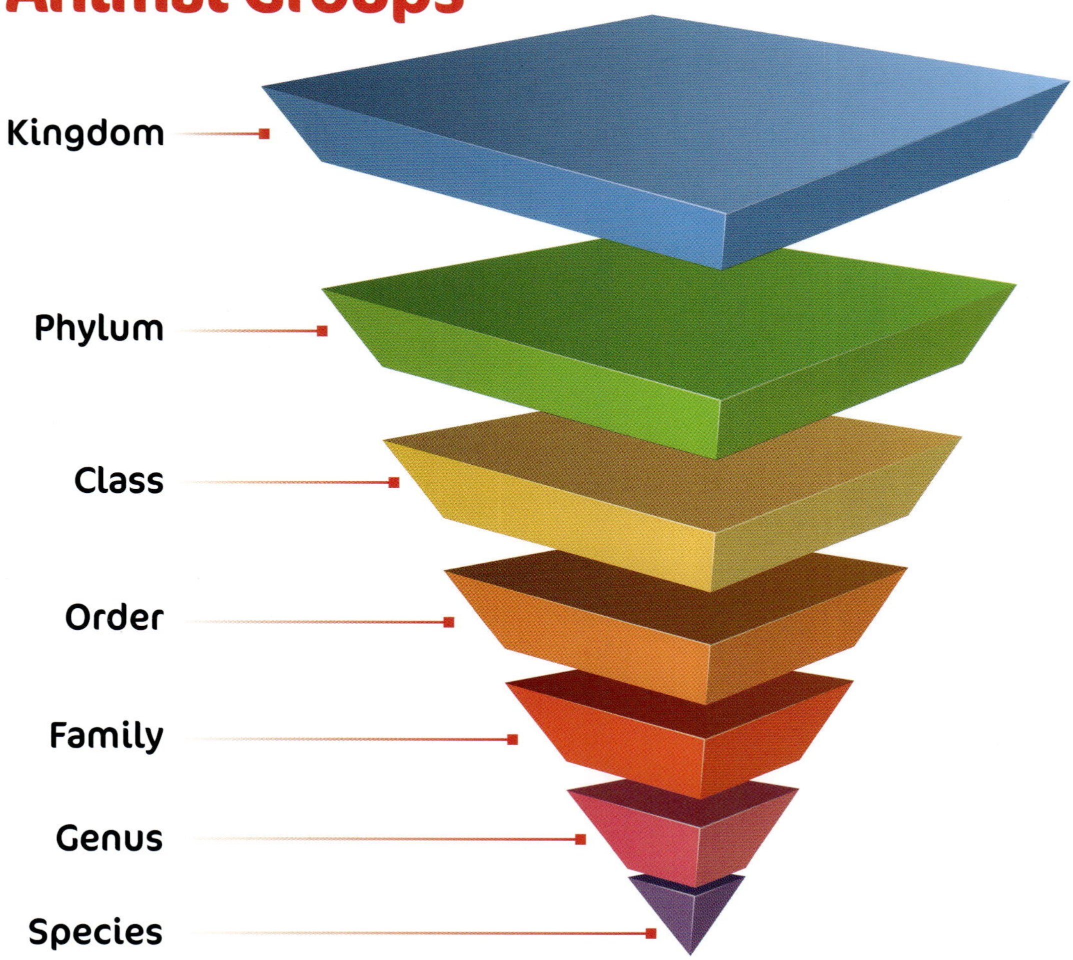

Kingdoms are broken into smaller groups called phyla. Mammals and other **vertebrates** belong to the phylum Chordata.

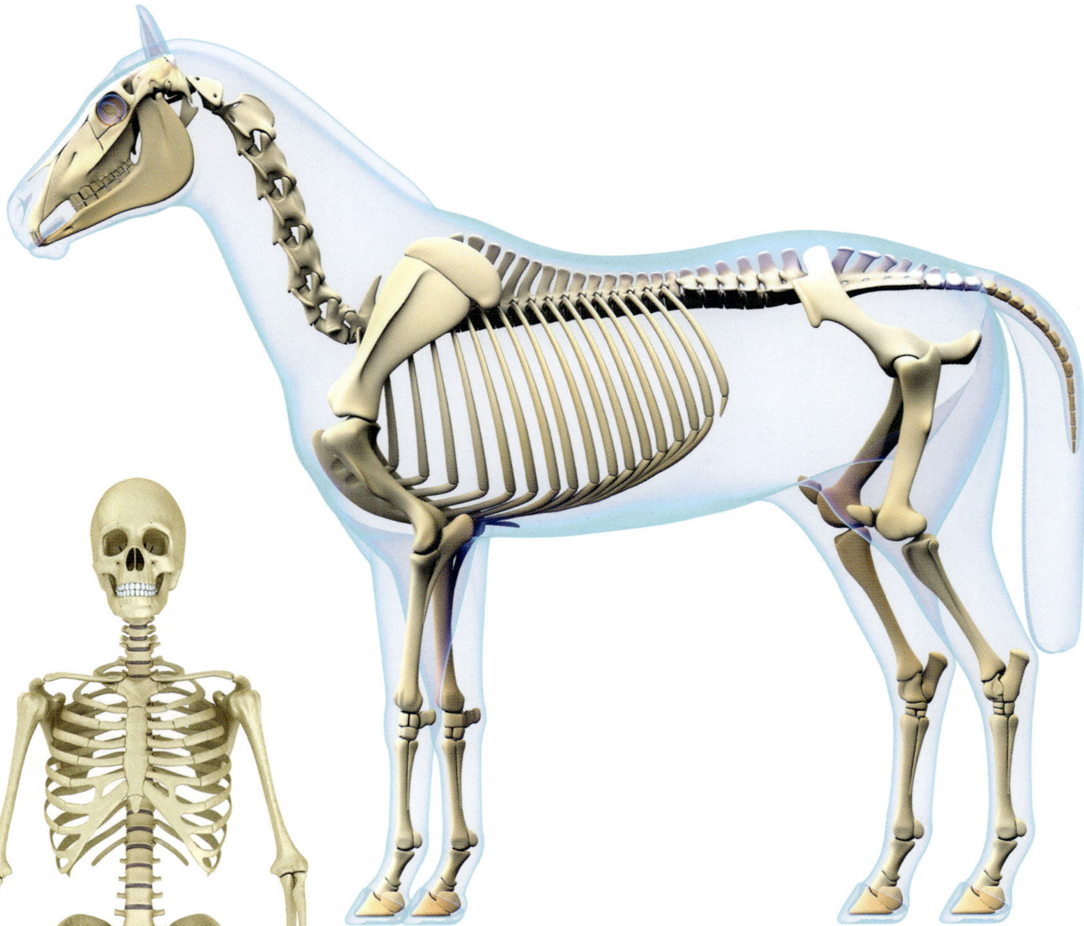

Horses have about 54 vertebrae depending on the breed.

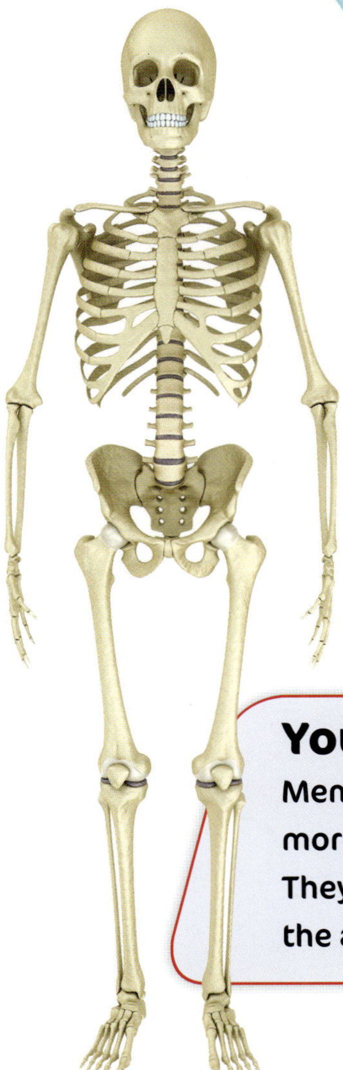

You're an Animal!

Members of the animal kingdom have bodies made of more than one cell. They cannot make their own food. They eat living or once-living things. People are part of the animal kingdom!

A phylum is broken into smaller groups called classes. Each class is made up of animals that share traits. These qualities set each class apart from all other animals.

Blue whales are the largest animals on Earth. These marine mammals can measure up to 98 feet (30 meters) long and weigh up to 190 tons (172 metric tons).

Mammals make up one class of animals. This class is called Mammalia. There are more than 5,000 **species** of mammals. All belong to one of 19 groups called orders.

Humpback whales are among the most endangered marine mammals. These amazing mammals communicate through song!

Sorting Animals into Species

Orders are made up of families. Families are broken down into genera. A genus is made up of one or more species. A species is one kind of animal.

All mammal species are warm-blooded. Mammals use energy from food to make body heat. All mammals have hair or fur. Most give birth to live young. Baby mammals drink their mothers' milk.

Female gorillas have a close bond with their young. Babies are dependent on their mothers for up to four years. Gorillas belong to the order Primates.

An American bison, or buffalo, cares for her calf on the grasslands. They belong to the order Artiodactyla.

Naming Mammals

Every species has a common name and a two-part scientific name. The first part tells which genus the mammal belongs in. It always starts with a capital letter. The second part tells the species name. It begins with a lowercase letter.

Home Sweet Home

Mammals live in many **habitats**. They can be found everywhere except the deepest ocean. Some have layers of fat to live on the ice. Others that live in deserts drink little water.

A mammal's habitat can help scientists group them into orders and families. Most species of mammals live on land. Some live in cold polar regions. Others live in steamy jungles.

Sea lions, seals, and walruses use their flippers to walk on land. They are called pinnipeds, which means "fin-footed" in Latin.

Marine mammals live in water. They have lungs to breathe air like land animals do. Most can stay under water for a long time. They return to the surface to breathe.

Dolphins, porpoises, and toothed whales never leave the water. They belong to the Delphinidae family in the order Cetacea.

Marine Mammal Groups

There are five groups of marine mammals: pinnipeds, cetaceans, sea otters, sirenians, and polar bears.

What's for Dinner?

Mammals can be grouped based on their diets. Herbivores only eat plants. Zebras nibble on grasses and leaves. White rhinos graze on grasses. These plant eaters live and travel in groups.

Some mammals eat mostly insects. They are built to dig for bugs. Their snouts are long. Their claws are sharp. Insect eaters have poor eyesight. Their other **senses** are strong.

Moles tunnel into soil. They dine on earthworms.

Traveling in groups helps plant eaters stay safe from predators.

Beavers build lodges from trees. They eat bark, leaves, and roots.

Carnivores eat meat. These animals are **predators**. They are built to hunt and eat prey. Omnivores eat plants and meat. They change their diets with the seasons. This helps them survive in many places.

Wolves hunt in packs. They catch big mammals such as deer.

Tigers hunt alone. These big cats kill buffalo and wild pigs.

Sea lions hunt fish and squid.

Mammal Babies

Mammals can be sorted into three groups based on how their babies are born. **Marsupials** are born early and grow inside their mothers' pouches.

Marsupial mothers give birth to tiny babies. The newborns climb into pouches. There they nurse on their mothers' milk and grow.

Virginia opossums are the only marsupials in North America. After babies leave the pouch they cling to their mother's back.

Kangaroos are only one inch (two centimeters) long when born. They stay in their mothers' pouches for up to ten months.

Placentals are the second group. They grow and change inside their mothers' bodies. As babies develop they get food and oxygen through the placenta.

Most mammals are placental. Some are usually born one at a time, such as humans and elephants. Others are born in a group, such as rabbits.

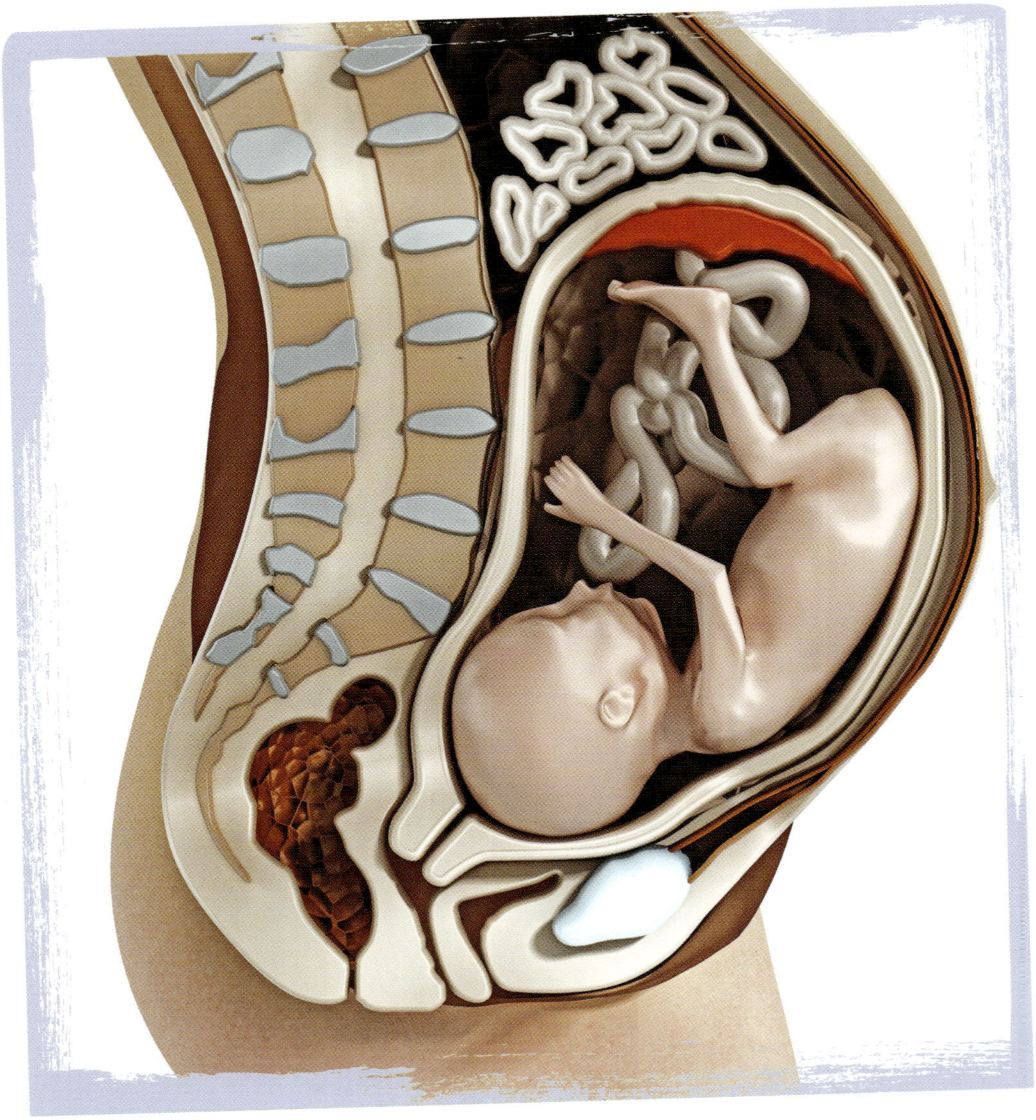

The placenta joins a human mother to her unborn baby. It supplies food and oxygen to the growing baby and removes waste products.

Monotremes are the third group. They hatch from eggs. There are only two species of monotremes: echidnas and platypuses.

Females lay leathery eggs that look like reptile eggs. After the eggs hatch, the babies drink their mothers' milk. The milk flows from **pores** on the mother's skin.

Echidnas and platypuses are only found in Australia.

The animal kingdom is huge! Scientists are still discovering new mammals. They also learn new things about known mammals. Sometimes new discoveries change the way species are classified.

Jessica Hale studies sea otters at the Olympic Coast National Marine Sanctuary in Washington State.

ACTIVITY

Design Your Own Mammal

Imagine you have discovered a new mammal. What does this animal look like? How is it similar to other mammals? How is it different? What will you name it?

For example, the new mammal might look like a mixture of a wolf and a bear. But the animal you discovered has hair instead of fur, is green rather than brown, and only eats insects.

Draw the animal and label the different traits that prove that it is a mammal. Then, use discarded materials such as cardboard and fabric scraps to create a 3D model of your mammal. Be creative!

How would you classify your new mammal based on what you know about animal classification?

Glossary

cells (SELS): the tiny units that together form all living things

habitats (HAB-uh-tats): the places where a plant or animal grows or lives in nature

marsupials (mahr-SOO-pee-uhls): mammals that carry their babies in a pouch on the mother's stomach

pores (POHRS): tiny openings (as in the skin)

predators (PRED-uh-ters): animals that obtain food mostly by killing and eating other animals

senses (SENTS-es): the five natural powers (touch, taste, smell, sight, and hearing) through which you receive information about the world around you

species (SPEE-sheez): a group of living things that ranks below a genus

vertebrates (VUR-tuh-brits): animals that have backbones extending down the back of their bodies

Index

Show What You Know

1. Why do scientists sort the animal kingdom into different classes?
2. Where do mammals live?
3. Why do many herbivores travel in groups?
4. What is unusual about how echidnas and platypuses babies develop?
5. Why do marsupial mothers need pouches for their babies?

Further Reading

Romero, Libby, *Ultimate Explorer Field Guide: Mammals*, National Geographic Children's Books, 2019.

Hall, Katharine, *Mammals: A Compare and Contrast Book*, Arbordale Publishing, 2016.

Sill, Cathryn and John, *About Mammals: A Guide for Children*, Peachtree Publishers 2014.

About the Author

Nancy Furstinger is the author of nearly 100 books, most featuring her favorite topic: animals! She shares her home with big dogs, house rabbits, and a chinchilla (all rescued), and volunteers with several animal organizations. You can learn more about Nancy and her books at www.nancyfurstinger.com.

Meet The Author!
www.meetREMauthors.com

www.rourkeeducationalmedia.com

PHOTO CREDITS: Cover, Title Page and Border ©Freder; Pg 3 ©aiqingwang; Pg 4 ©lvcandy; Pg 5 ©decade3d, ©alex-mit; Pg 6 ©bingokid, ©Anthony Lombardi; Pg 7 ©blake81; Pg 8 ©joebelanger; Pg 9 ©mlharing; Pg 10 ©GlobalP; Pg 11 ©lemga, ©GlobalP; Pg 12 ©Tramper2; Pg 13 ©mlharing, ©Bob Hilscher; Pg 14 ©Lynn_Bystrom, ©labrlo; Pg 15 ©RobsonAbbott; Pg 16 ©Specialjake; Pg 17 ©mburt; Pg 18 ©adventtr; Pg 19 ©JohnCarnemolla, ©bennymarty; Pg 20 ©Kate Thompson/NOAA; Pg 22 ©kickers

Edited by: Keli Sipperley
Cover and interior design by: Kathy Walsh

Library of Congress PCN Data

Mammals / Nancy Furstinger
(Animals Have Classes Too!)
ISBN 978-1-64369-028-5 (hard cover)
ISBN 978-1-64369-110-7 (soft cover)
ISBN 978-1-64369-175-6 (e-Book)
Library of Congress Control Number: 2018956023

Rourke Educational Media
Printed in the United States of America,
North Mankato, Minnesota